The
Geddy Foundry

A Colonial Williamsburg Historic Trades Book

The Geddy Foundry

By
Sven Dan Berg and
George Hassell

The Colonial Williamsburg Foundation
Williamsburg, Virginia

Library of Congress Cataloging-in-Publication Data

Berg, Sven Dan, 1936–
    The Geddy foundry/by Sven Dan Berg and George Hassell.
        p. cm.—(A Colonial Williamsburg historic trades book)
    Includes bibliographical references.
    ISBN 0-87935-086-5
      1. Foundries—Virginia—Williamsburg—History—18th century.
    2. Geddy, James, d. 1744. 3. Williamsburg (Va.)—Industries.
    4. Metal-work—Virginia—Williamsburg—History—18th century.
    I. Hassell, George, 1946– . II. Title. III. Series.
    TS229.5.U6B47 1991
    671.2′09755′4252—dc20   91-19034
                                       CIP

Designed by Helen Mageras

Cover photography by David M. Doody

Printed and bound in Singapore

Preface

We offer this little book as a brief introduction to one of the trades practiced in eighteenth-century Williamsburg. It is not a detailed technical treatise nor an in-depth socioeconomic study. Maybe it would be best described as reflections of two craftsmen on the nature of their craft and on the lives of some people who practiced it before them.

Both authors are brass founders for Colonial Williamsburg, products of a craft preservation effort that is without equal anywhere in the world. We have worked with one another at molding bench and forge for a considerable time; first as master and apprentice, then as master and journeyman. One way or another we have been in harness together for more than twenty years.

Our essay focuses on a single family: the Geddys of Williamsburg. We are fortunate in having more information about the Geddys than about most other craftspeople in colonial Virginia. It is not a lot — not as much as we would like to have — but enough for at least a partial portrait of a family making their way in the world with skilled hands and honest industry.

We hope our approach to our topic will give some suggestion of how close and complex is the relationship between technical and social history. We also hope that we can share with the reader something of the bond that we feel with the people who lived and labored here two centuries ago.

*T*he first printed book on practical metalwork was published in Venice in 1540. It was titled *Pirotechnia* and was written by a craftsman, the Sienese smith and foundryman Vannoccio Biringuccio. Within this early handbook of metallurgical technique, Biringuccio included some general observations on the nature of the founder's art: "Therefore, having considered this work many times, with its extraordinary obstacles and the bodily labors heavy as a stevedore's, instead of exalting it with praise, I wish to say that it is such that a man of noble birth . . . should not practice it and could not unless he is accustomed to the sweat and many discomforts which it brings. . . . Nor do I doubt that whoever considers this art well will fail to recognize a certain brutishness in it. . . . In addition, this art holds the mind of the artificer in suspense and fear regarding its outcome and keeps his spirit disturbed and almost continually anxious. For this reason they are called fanatics and are despised as fools. But, with all this, it is a profitable and skillful art and in large part delightful."[1]

Speaking of a brass foundry he visited in Milan, Biringuccio noted that "whoever entered that shop . . . would, I think, believe as I did that he had entered an Inferno, nay, on the contrary, a Paradise, where there was a mirror in which sparkled all the beauty of genius and the power of art."[2]

Although Biringuccio wrote in the middle of the sixteenth century, his characterization of foundry work would surely have been recognized and approved as essentially accurate by those who practiced the trade in Williamsburg in the eighteenth century. Though there had been significant technological innovation in the intervening two hundred years, the general nature of the work had remained much the same. Perhaps even less change occurred in the attitude of the craftsman toward his work, which Biringuccio so vividly expressed: a realistic appraisal of its exhausting demands in time, labor, and emotional energy combined with a keen appreciation of its potential rewards, in terms not only of economics but of personal satisfaction as well.

It would be difficult to overstate the influence of the pre-industrial craftsman's trade on nearly every aspect of his life. The acquisition of the requisite technical knowledge and manual skills was almost the sole object of an educational process that began in childhood and took place almost entirely in the workshop. The skills so acquired were often the craftsman's principal inheritance and gave him a place in society that was not dependent on birth or possession of landed property. The practice of the trade was the principal occupation of the vast majority of his waking hours and was the

---

A crucible of molten brass ready for pouring is lifted carefully from the forge.

ultimate source of all the wealth, position, and pride that he and his family might claim. It is not surprising, then, nor inappropriate, that with crafts families the record of their lives that we are able to piece together seems to so great a degree to be the record of their business in the world.

The Geddy Foundry Of Williamsburg

On August 8, 1751, the following advertisement appeared in the *Virginia Gazette*: "DAVID and *William Geddy* Smiths in *Williamsburg,* near the Church, having all Manner of Utensils requisite, carry on the Gunsmith's, Cutler's, and Founder's Trade, at whose Shop may be had the following Work, *viz.* Gun Work, such as Guns and Pistols Stocks, plain or neatly varnished, Locks and Mountings, Barrels blued, bored, and rifled; Founder's Work, and Harness Buckles, Coach Knobs, Hinges, Squares, Nails and Bullions, curious Brass Fenders and Fire Dogs, House Bells of all Sizes, Dials calculated to any Latitude; Cutler's Work, as Razors, Lancets, Shears, and Surgeon's Instruments ground, cleaned, and glazed, as well as when first made, Sword Blades polished, blued, and gilt in the neatest Manner, Scabbards for Swords, Needles and Sights for Surveyors Compasses, Rupture Bands of different Sorts, particularly a Sort which gives admirable Ease in all Kinds of Ruptures: Likewise at the said Shop may be had a Vermifuge, Price, 3s. 6d. *per* Bottle, which safely and effectually destroys all Kinds of Worms in Horses, the most inveterate Pole-evils and Fistulas cured, and all Diseases incident to Horses; at their said Shop."[3]

At the time this advertisement appeared, the citizens of Williamsburg

Left: *The Geddy Foundry at Colonial Williamsburg.*
Above: *David and William Geddy advertised their trade in Hunter's* Virginia Gazette, *August 8, 1751.*

9

would already have been familiar with the shop "near the church," for the Geddy family had lived and worked at metal crafts on this lot on Palace green for nearly fifteen years, and the next quarter-century would see a third generation of the family apprenticed to the trade on this site.

Probably the thing that first strikes the modern reader of David and William's advertisement (apart from the sideline business in trusses and vermifuge) is the great variety of trades that are represented. The possession of this rather astonishing range of skills seems not to have been peculiar to the Geddys but was characteristic of many other American craftsmen as well, though perhaps to a somewhat lesser degree. Diversification, in fact, was the rule rather than the exception among tradesmen in eighteenth-century Virginia.

David and William grouped the work they offered into three broad categories: "the Gun-smith's, Cutler's, and Founder's Trade." Gunsmithing, of course, involved the making and repair of firearms. Cutlery, as the advertisement shows, included the production of cutting tools ranging from swords to surgeons' instruments. The relationship of the objects included in "Founder's Work" seems less obvious at first glance: what do a house bell, a firedog, and a harness buckle have in common? The answer lies not in any similar function of the articles but rather in a common method of fabrication: they are all things that are formed by casting.

In the eighteenth century there were two basic ways of shaping metal: hammering it, which was generally the work of a smith; and casting it, the process of melting metal and pouring it into a mold, which was the work of the founder. There was a natural tendency for softer, more malleable metals to be hammered and for harder or more brittle metals, or those that are particularly fluid, to be cast. But perhaps an even more important factor would be the shape of the article to be made. A bowl, a tray, or a cup would generally be hammered; a candlestick, a bell, a shoe buckle, a watch key, or a spandrel for a clock would be cast shapes. These and hundreds of similar articles in a variety of metals would have been produced in the foundry that the Geddy family operated for nearly fifty years.

The Development Of Casting

The founder's trade is an ancient one, extending back at least five thousand years. It is likely that the earliest casting process employed open stone molds. A cavity would have been hewn into the rock surface and the

A collection of cast objects including candlesticks, buttons, a marrow scoop, serving spoons, and a pipe tamper.

molten metal then poured into it to produce a rough shape, which was given final form under the hammer. Though this process was perfectly suitable for producing a copper knife or other simple objects, it was extremely limited in its application to the production of more complex forms. A major breakthrough occurred with the concept of indirect molding, using a model or pattern to create an impression in clay into which the metal would then be poured. By employing sectional molds formed in this manner, more complicated shapes such as socketed axes could be cast, with considerable improvement also in surface finish and detail. Because the pattern itself was permanent, identical castings could be produced at will.

Sometime during the Bronze Age the technique of lost wax casting was introduced into the Mediterranean cultures. In this process a pattern of wax is formed and packed in clay, or invested, to form a mold. The clay mold is then heated to melt out the wax, leaving the cavity into which the metal is poured. This method was used particularly for casting large, complex shapes such as statuary.

For the next three thousand years virtually all high-temperature metal casting (in Western cultures at least) was done with either clay piece molds or the lost wax technique. Many cast works surviving from antiquity demonstrate an extremely high level of technical proficiency. However, for several centuries after the fall of Rome, casting technology in Europe seems to have suffered the same partial eclipse as virtually all other arts and sciences. It was not until the late Middle Ages that development seems to have resumed, partly with the rediscovery or renewed practice of ancient casting processes, partly with the introduction of altogether new techniques, of which the most significant was sand casting.

This new process of sand casting was probably an outgrowth of the clay piece-molding technique since, as with that method, a permanent pattern was used and the mold was formed in sections to allow the removal of the pattern. The material of the mold, however, was not clay but an extremely fine sand with a slight admixture of clay or some other material to act as a binder. The entire mold was formed within a sectional wood or metal frame called a flask. In versatility, speed of production, and quality of cast surface, sand casting eventually proved superior to any of the ancient casting techniques for many types of work. Nonetheless, the new technology was not immediately and universally adopted throughout the trade. Although Leonardo da Vinci and other Renaissance artisans were familiar with the technique, it was not really until the late seventeenth and early eighteenth

Battery works were involved in producing sheet metal and wire, both stock material and finished goods. "L'Art de Convertir le Cuivre Rouge," Descriptions des Arts et des Métiers, *Vol. XI (1767; Slatkine Reprints, Geneva, 1984), plate 18.*

centuries that sand casting assumed a dominant position in the foundries of Europe. Its widespread introduction and rapid development during this period undoubtedly played a key role in the tremendous expansion in the foundry industry that occurred over the next hundred years, perhaps nowhere more dramatically than in England.

Development Of The English Brass Foundry Industry

The English brass industry in its early stages included few foundries. It consisted principally of battery works, which were involved with the production of sheet metal and wire—both stock material and finished goods—by a process of hammering, as the term "battery works" would suggest, or, in the case of wire, by drawing the metal through dies. One of the major products of this period was brass wire for wool cards that were needed in great quantities by England's flourishing wool industry.

By the end of the seventeenth century, however, more foundries began to appear, and their number multiplied rapidly throughout the eighteenth century. The city of Birmingham in the Midlands had seventy-one brass

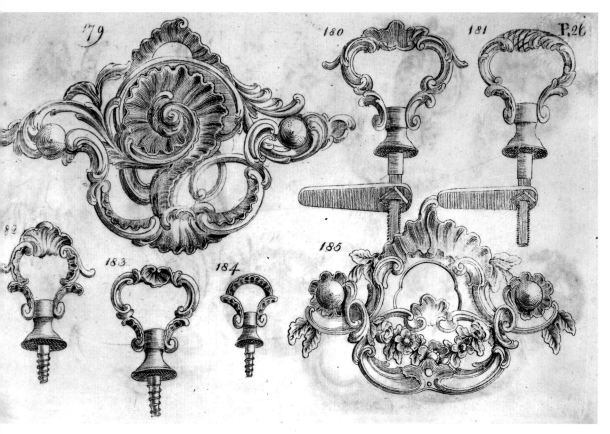

foundries by the end of the century.[4] Birmingham was, in fact, something of a center for the brass industry in England. There were certainly other cities with numerous foundries, but Birmingham presented the most striking example of a concentration of similar manufactories producing for a national and international rather than a local market.

As might be expected, there tended to be a progressive specialization in these shops as the century wore on: some foundries began to concentrate on the production of candlesticks, furniture brasses, or fittings for carriages and harnesses. These products were distributed to retail shops through independent wholesalers, or factors, who traveled throughout England with pattern books, collections of full-size engravings of the available products identified only by numbers in order to keep the identity of the makers known only to the factor, thereby preventing the retailer from dealing directly with the founder. In addition to this national market there was also a considerable export trade in English brassware. Many of the stores in

A page from a pattern book. Notice the identifying numbers. Sample Book 739 4 B82, Brass Ornaments for Furniture, Pattern-book, [1770?], p. 26. Courtesy of the Essex Institute, Salem, Massachusetts.

Williamsburg sold the products of the Birmingham foundries.

In general, by the third quarter of the eighteenth century, the English brass trades formed a large-scale and complex industrial organization. It included mining and smelting operations, firms that alloyed the brass and produced partly fabricated material (wire and sheet for the braziers to hammer, ingot for the foundries to cast), the makers of the finished product (many of them specialized), a distribution network of factors and exporters, and, finally, the retail outlets.

Looking again at William and David Geddy's advertisement, it seems apparent that the nature of their business was formed by market forces or trade practices that differed significantly from those in England. To see this more clearly it might be useful to look briefly at the development of the Geddy family's enterprises.

Growth Of The Geddy Family's Businesses

The first of the Geddy family to locate in Williamsburg was James Geddy, Sr. It is not known for certain when he arrived in Virginia or where he came from. There is some reason to suppose that he emigrated from Scotland as a young man, but the evidence is not conclusive, and it is possible that he was born in the colony. In any case, he appeared in Virginia court records by 1733 and was living in Williamsburg at the corner of Duke of Gloucester Street and Palace green at least by 1738. Two advertisements that he placed in the *Virginia Gazette* have survived. They show that he was doing gunsmithing, wrought brass work, and the casting of small bells.[5]

Geddy seems to have been reasonably successful in his business affairs: he ultimately left an estate valued at £178, about average for a craftsman of his time.[6] He may, in fact, have been so busy that he found it inconvenient to attend his parish church with the regularity required by law at that time, even though the church was located on the other corner of Palace green and Duke of Gloucester Street, directly opposite his own establishment. The records of the May 1741 session of the York County Court show a grand jury making a presentment against "James Geddy; For not going to his parish Church." The record of the following month's session shows James Geddy also not going to his county court, having failed to appear to answer the charge. In the July session he was finally ordered to pay a fine of five shillings or fifty pounds of tobacco.[7]

James and his wife, Anne, had eight children, four boys and four girls.

Apparently he intended that his sons follow him in the metal trades: his two older boys, William and David, were serving apprenticeships under their father when the family situation changed abruptly in late summer 1744. Court records provide two notices of James Geddy at this period: in July he was a member of a grand jury in Yorktown; in August his will was probated.[8] The Geddy family was suddenly left without its head and the Geddy family's business without its master.

Fortunately, Anne Geddy seems to have been a strong and capable person. There is some evidence that her husband regarded her as such since he willed his entire estate to her, a somewhat unusual occurrence for that period. It is likely that Anne was able to assume direction of the business affairs of the family. In October of the same year she petitioned the House of Burgesses for the payment of a debt owed to her deceased husband for cleaning seven hundred weapons in the Magazine by order of the governor. Anne's petition was initially rejected, but, undeterred by this setback, she persisted in her application and was finally awarded a payment of £21 8s. 4d.—a substantial amount of money. Anne's two older sons, David and William, were probably sufficiently skilled at the time of their father's death to be able to continue the operation of the foundry and gunsmith shop, perhaps with the assistance of an indentured servant named William Beadle, who is mentioned in the inventory of their father's estate and who may well have been a skilled craftsman.[9]

The case of Adam Pavey gives some indication of the difficulties that the death of a shop's master could create. Pavey, a blacksmith and pewterer in Spotsylvania County, Virginia, had an indentured servant named William Ashwell who worked for him as a brass founder. Pavey died in June 1756, at which point Ashwell still had a year of his indenture to serve. The minute book of the Spotsylvania County Court on July 6, 1756, records a "petn [petition] of Wm Ashwell against his mistress Agness Pavey for missvage [misusage] ordered that the sd Agness be summoned to the next court to answer the sd complt."[10] The exact nature of the disagreement and the outcome of the legal action remain unknown, but this brief notice is sufficient to suggest a definite personnel problem, as it might be termed today. Anne Geddy seems to have been more successful in providing a smooth transition through what must have been a rather difficult period.

The situation of Anne's two younger sons, however, was rendered somewhat uncertain by their father's death. James, Jr., was only thirteen at this time and John even younger, so neither could have progressed very far

---

Checking the metal. Pouring temperatures must be judged by visual observation.

in an apprenticeship to his father. Consequently, Anne was also faced with the problem of providing for the future of these two younger boys. Although documentary evidence of the Geddys' activities during this period is extremely scant, there is some reason to believe that Anne put together an arrangement that killed two birds with one stone. By renting part of her property to Samuel Galt, a local silversmith, she secured extra income for the family. At the same time, by apprenticing the two younger boys to Galt, she provided for their education. This speculative scenario is based on the knowledge that Galt worked on the Geddy site at one point in his career and that James and John Geddy both did become silversmiths.[11]

While James and John were serving their apprenticeships, William and David continued to pursue their father's trades. During 1748–1749, however, David left the family business to attempt to set up on his own as a blacksmith in Fredericksburg, Virginia. The court records from that city indicate that David's business quickly failed under the weight of a rapidly accumulating number of suits for debt brought against him by local citizens.[12] William may have assisted David in extricating himself from his financial difficulties and returning to the family business, for by 1751, the year of the *Virginia Gazette* advertisement, David was back in partnership with brother William.

A crucible in the forge. The entire course of the melt lasts an hour or more.

At about this time young James would have been finishing his apprenticeship, and John his shortly thereafter. John eventually left Williamsburg and set up shop in Halifax, North Carolina. James remained in town working as a silversmith and apparently prospering, for in 1760 he purchased the house and lot on Palace green from his mother and established his silversmithing and jewelry business on the property. A few years later he tore down the house in which he had been raised and built the larger structure that still stands on the site.[13] Following the family tradition, James apprenticed his two sons to himself and also brought his brother-in-law, William Waddill, into the business as a watchmaker and engraver. David and William continued to operate the foundry as a separate enterprise, though they probably paid their brother rent after his purchase of the property.

William's Foundry

By the 1760s, then, the original shop of James Geddy, Sr., had expanded to a sort of family conglomerate with perhaps as many as twelve to fifteen craftsmen and apprentices working at seven or eight different crafts and with at least two distinct businesses in operation on the site. Even within the foundry itself the scope of the work was broadening. William was still making guns and casting bells as his father had, but cutlery work and general blacksmithing were now being done, and the range of the cast work had been expanded to the point where it had probably become the principal activity of the shop. Certainly the archaeological investigation of the Geddy site turned up more evidence of brass foundry work than of any other of the crafts practiced here.

As was mentioned earlier, the fact that William and David were practicing a variety of trades was not unusual for their time and place. Just two months before David and William ran their advertisement in the *Gazette,* one Ephraim Goosley of Yorktown advertised in the same paper that "Gentlemen, and Others, that have Occasion of any Kind of Iron or Brass Work, either polish'd or rough, may be supply'd on applying to the Subscriber" and that furthermore "Gun-work, such as new Stocks, Cocks, Mounting, etc. are done after the best Manner." From a slightly later period the advertisement of James Haldane, formerly of Philadelphia and Norfolk, read, "now settled in *Petersburg* [Va.], where he carries on the COPPERSMITH's, BRASS-FOUNDER's, TINNER's, PEWTERER's, and PLUMBER's Branches."[14]

Diversification offered definite advantages to the colonial tradesman. The key factor was the extremely rapid growth of the Virginia colony and the relative shortage of skilled labor. Large numbers of new households were constantly being established as the population doubled every twenty to twenty-five years, creating a great demand for all sorts of metal goods. Whereas in Birmingham a heavy concentration of craftsmen was dependent on a distribution system of factors to reach their market, here in Virginia a scarcity of craftsmen worked in a large and rapidly growing local market. Craftsmen here found that possessing and practicing a wider range of skills opened up larger areas of this profitable retail trade, an advantage that greatly outweighed any loss in efficiency of production. It was not unusual for the master of a shop to hire specialists to expand the range of his business. As seen earlier, James Geddy, Jr., brought William Waddill, a watchmaker and engraver, into his business, and the brass founder William Ashwell worked for Adam Pavey, a blacksmith and pewterer. In 1773 Peter Hardy, a Williamsburg coachmaker, advertised that "Having lately provided himself with an extraordinary good Workman in the BRASS FOUNDRY Business (whose knowledge in that Art he would venture to

A small works foundry: Fig. 1. The forge, with a half-mold leaning against the wall being dried. Fig. 2. Craftsman pumping bellows. Behind him a mold being dried over a brazier. Fig. 3. Making a sand mold. Fig. 4. Pouring metal. Fig. 5. Empty flasks. Crucibles on shelf above. Denis Diderot, Encyclopédie, Recueil de Planches, sur les Sciences, les Arts Liberaux, et les Arts Méchaniques, Avec Leur Explication, *Vol. V (Paris, 1767),* Fondeur en Sable, *plate 1.*

affirm is equal to any in *America,* if not superior) the Public may be supplied with all Kinds of BRASS WORK upon the most reasonable Terms."[15]

It must be remembered, however, that the colonial craftsmen did not have the local market entirely to themselves. Directly across Duke of Gloucester Street from the Geddy property, John Greenhow's general store offered for sale a wide range of English-made brass and other metal wares, as did many other stores in Williamsburg and throughout Virginia. The original production cost of these goods was probably less than that for comparable colonial work since the English shop was often a relatively high-volume and specialized producer and since skilled labor in England was cheaper than in America.

But the cost of transport and the mark-up of the merchant increased the price of imports. Furthermore, the supply could be uncertain; a store might run out of a given article and not be able to replace it immediately. Consequently, there was still a market for the more expensive locally produced goods, and someone like the aforementioned Ephraim Goosley of Yorktown could confidently advertise "all Sorts of Axes and Hoes, at the low Price of Forty *per Cent* more than they cost in *London.*"[16] Of course, nothing prevented the local craftsman from claiming a part of the import business for himself, and this seems to have been done often. In the case of the Geddys, James, Jr., the silversmith, also ran a retail shop and offered for sale a great variety of imported English luxury goods. Moreover, virtually all the repair work business would necessarily have fallen to the local tradesman, and this seems to have provided a substantial portion of the colonial craftsman's income.

In one area the colonial craftsmen who worked in nonferrous metals seem to have operated at some disadvantage: in obtaining the raw materials of their trade. Raw, newly refined metal was not readily available since, except for iron, very little mining or smelting of metals was done in Virginia or any of the other colonies at this time, and there were legal restrictions on the export of unworked metals from England. The craftsmen instead relied on recycling—buying up broken or worn-out metal goods, sometimes provided by the customers themselves. Advertisements of metalworkers in Virginia almost always included some offer to purchase old metal at favorable terms.

One good source of metal for the Geddy foundry was Alexander Craig, a local harnessmaker and tanner. One of Craig's daybooks for the period 1761–1763 survives. It contains frequent entries recording Craig's sale of

Weighing scrap metal. Raw, newly refined metal was not readily available in eighteenth-century Virginia.

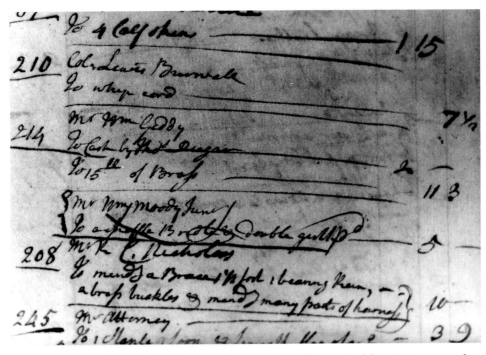

scrap brass, probably old harness brass, to William Geddy. Craig was also selling brass to William Ashwell, who had apparently completed that last, onerous year of indentured servitude to the Widow Pavey and had set up business in Williamsburg. It is interesting to note that no entries record Craig purchasing new metalwork from Geddy, but on two occasions he did buy substantial amounts of harness brass from Ashwell. Since this daybook covers only a brief period, however, it is not possible to draw any conclusions about Craig's preferences in suppliers of brass for his harness business. Occasionally, craftsmen may have sought scrap material from sources outside the colony, for Goosley of Yorktown imported "junk"—presumably scrap metal—from the West Indies.

Most other necessary tools and supplies were available in local stores. John Greenhow's store across the street from the Geddys carried molding sand, crucibles, polishing abrasives, fluxes for metals, and such, though the outbreak of war created shortages in some necessary commodities. In 1776 James Haldane of Petersburg not only advertised for old copper, brass, pewter, and lead but also said he was "in great Want of Spelter, or Spelter Soler, also Borax and Sal-Ammoniac (for which he will give the highest price) and coarse and smooth Files."[17]

Left: A crucible full of scrap brass. Above: An entry in Alexander Craig's Account Book, 1761–1763, Galt Papers, April 13, 1763, shows that Thomas Dugar purchased metal from Craig for William Geddy. Courtesy, Earl Gregg Swem Library, College of William and Mary, Williamsburg, Virginia.

Blisland Parish   Dr

	£r. Tob.
To Debt brought forward	23997
To Sarah Foxe, for keeping ~~......~~ Joseph Foxes two Children	750
To David Allen, for keeping William Charlton till June 6. 1763	533
To Robert Buchan, in Maj: William Armistead's hands, for keeping Betty Kitson, till 1763, & burying her	600
To Capt: William Macon, for keeping Cliff Dugar	500
To Do for two Communions for the upper Church	250
To William Geddy, for keeping Thomas Dugar, who is to be bound to the P. Geddy, & Geddy to clear him from further Charge to the Parish	600
To Richard Cced Curle, a further allowance towards his Maintainance	400
To Majr. Thruston James, for keeping David Donner	1500
To Mary Jones, in Majr. William Armistead's hands	800
To Joel Willis, for keeping Thomas Twins Child	500
To David Hazelwood, for keeping Eliz.a Dunkerton	1000
To Margt. Manning, for keeping two of Sackvile Mhone's Children	800
To Edward Lively	800
To Ann Taylor, (in Majr. William Armistead's hands) for the Support of her Son John Taylor	600
To Mr. William Smith, D. Shord. & Collector, his acct.	395
To Capt. Edward Power, for two Communions, for the lower Church	250
	34475
To 6 p.C. for Collecting 34475 Tob.	2068
	36543

Blisland Parish   Cr

	£r. Tob.
By 1100 Tithes at 34 Tob p poll	37400
The Debt as above, is	36543
Deposited in the Collector's hands	857 Tob. to be paid to the Church wardens towards repairs on the Glebe

N.B. The Cash Accot. & Orders Carried over to the other Side

With the growth of the Geddys' business, there was an accompanying increase in the number of craftsmen working on the site. By the 1760s the two businesses—silversmithing and foundry—together may have employed a dozen or more craftsmen: masters, journeymen, and apprentices. The Geddy foundry had two masters, or owners, William and David. Actually, David's presence on the site is somewhat problematic since he does not appear in any records between 1751 and the outbreak of the Revolution. Apparently David enlisted in the Revolutionary forces and served in arms, but the course of his subsequent career or, for that matter, whether he even survived the conflict is not known.[18] This scarcity of documentary evidence relating to David, added to the record of the failure of his Fredericksburg enterprise, encourages the suspicion that William was the dominant figure in the partnership, being gifted, perhaps, with superior business sense.

In our admiration for the remarkable technical skills so often displayed by the eighteenth-century artisan, it is easy to forget that the possession of these skills did not necessarily ensure success. The craftsman—at least the master craftsman—was of necessity also a businessman, responsible for the acquisition of materials, the marketing of the product, the efficient organization and assignment of the daily work load, and the management of what could be a rather heterogeneous work force. The labors of the master in the counting room could determine the success or failure of the business quite as much as his labors at the forge.

That William possessed the necessary skills seems indisputable. He preserved the business of his father, expanded it, and operated it successfully for over thirty years. He himself achieved a comfortable level of prosperity. It is unlikely that William lived on the property where he worked, the house on this lot being occupied by his brother James and his family. Apparently he rented a residence in town; city tax rolls record his payment of personal property taxes on himself, four slaves, three cows, and two horses. In addition, he owned a 326-acre farm about ten miles outside of the city, and at some point he erected on this property a large three-story frame house that is still standing. Anne Geddy, incidentally, also owned 100 acres in this area, perhaps adjacent to William's tract, and was living there in what might be thought of as a well-earned and contented retirement, having seen all her sons finally established in their businesses and communities.[19]

William had at least two apprentices. One, following the Geddy pattern, was his own son, William, Jr. The other was an orphan boy named Thomas Dugar, who had been apprenticed out to William by the vestry of

Blisland Parish Church, which was near William's farm.[20] County courts or parish vestries often provided for orphans in this fashion. It is not known how old Dugar was at the time of his apprenticeship, but orphans were sometimes apprenticed when only six or seven years of age, which meant that if they served until the age of twenty-one—the normal procedure—their apprenticeships could last as long as fifteen years. The typical apprentice, of course, started at a somewhat later age, say around thirteen or fourteen, but the relative youth of a large portion of the work force in any eighteenth-century shop would be striking to a twentieth-century observer and probably created certain problems in employee relations not normally encountered in the modern place of business.

William employed an older man, John Dennis, as a journeyman, a skilled craftsman working for wages. Dennis brought with him to the shop a considerable fund of metalworking experience, having been employed in the trade at least since 1741, when he was drawing journeyman's pay at the Tayloe Ironworks on the Rappahannock River. It is interesting to see that Dennis followed the typical colonial pattern of diversification of skills, starting out at an iron furnace, moving to a brass foundry, and leaving in his inventory when he died the tools for the gunsmithing and silversmithing trades. Dennis was also typical in that, after leaving Geddy's employ, he established his own business in Halifax County, Virginia.[21]

Virtually all craftsmen in colonial Virginia made that step from wage earner to shop owner, in marked contrast to the situation in England where many journeymen remained in that position for their entire careers, never becoming independent masters. The great opportunities that existed for craftsmen in the rapidly expanding colonial economy might help to explain the absence in the colony of any indication of the sort of labor conflict that began to appear during that period in England, where journeymen crafts-men in numerous trades formed temporary combinations to protest wages or working conditions. Since nearly all colonial journeymen could reason-ably expect to become masters of shops, usually within a very few years, they tended not to view themselves as a distinct and permanent class of workers whose economic interests were necessarily opposed to those of the masters.

One type of craft worker, however, did indeed belong to a well-defined and permanent class; his expectations were limited absolutely by factors other than his own skill and industry, and for him the great opportunities of the booming colonial economy did not exist. The use of slaves in the craft shops of Williamsburg seems to have been less common than might be expected, their labor as agricultural workers or domestics being perhaps of greater or more immediate value. But there were few trades in which slaves were not at least occasionally employed as skilled workers. Whether William Geddy used slave labor in his shop is not known. It is certainly possible, and the likelihood of it seems greater in light of the fact that William Geddy, Jr., did own a slave named Charles who was a blacksmith.

Since the average eighteenth-century citizen left no documentary evidence of his existence, it is likely that there were other workers in William Geddy's foundry who were not recorded. Those mentioned, though, are sufficient to represent the spectrum of people who would have been present in the shop.

Molding In Sand

"For the founder is always like a chimney sweep, covered with charcoal and distasteful sooty smoke, his clothing dusty and half burned by the fire, his hands and face all plastered with soft muddy earth."[22]

This is another description from Biringuccio's *Pirotechnia,* and like the ones cited earlier it is vividly phrased yet essentially accurate. Heat and dirt: perhaps the dominant impression of any visitor to a foundry now or in centuries past. The need for heat is obvious in a process where metals must be reduced to a molten state to work them, many of the metals the Geddys worked with requiring temperatures of 2000° F or higher. The dirt in a foundry, though, has a significance rather different from that of other trades. Part of it may be just dirt, a by-product of the work. But in a foundry, dirt—of a type—is also one of the essential tools of the trade.

When metal is cast at 2000° F, the mold into which it is poured must be composed of a material that is heat resistant, that is permeable to the gases generated in the casting process, and that can be formed with reasonable speed into the desired shape and will then hold that shape under the flow

Molders. Denis Diderot, Encyclopédie, Recueil de Planches, sur les Sciences, les Arts Liberaux, et les Arts Méchaniques, Avec Leur Explication, *Vol. IV (Paris, 1765), Forges, 3rd Section, plate 5.*

This illustration from Diderot's encyclopedia pictures young apprentices at work in a foundry. Diderot, Encyclopédie, IV, plate 9.

Fig. 5.

Fig. 8.

of the metal. As seen earlier, the most widely used molding material by the beginning of the eighteenth century was sand of specific types. Different kinds of cast work require different kinds of sand. There is, after all, considerable difference between casting a cannon and casting a shoe buckle. The sand that the Geddys and other small works founders used was described in the first edition of the *Encyclopaedia Britannica* as "of a pretty soft, yellowish and clammy nature."[23] Today it might be described as an exceptionally fine-grained silicate sand with a 4- to 5-percent content of alumina or clay, usually occurring as a sedimentary deposit on river beds or banks. It does indeed have a distinctive yellowish color and a soft, almost fluffy consistency when properly prepared. It is activated, or tempered, by the addition of a small but critical amount of water, which causes the clay to liquefy, coat the sand grains, and act as a binder.

Considerable labor, apprentice labor usually, is expended in preparing the sand for molding. In order to break up any lumps and distribute the moisture evenly, the sand is subjected to a process known as tewing. It is placed on a board and worked with a spatula-like implement, being alternately spread out with a smearing action, scraped back into a heap, and respread until it reaches a uniform and proper consistency. At the right state it is moist enough to retain a shape but is not sticky to the touch.

The sand does not have sufficient binding strength to stand completely unsupported. Consequently, the mold must be formed within a rectangular frame called a flask. The flask actually consists of two frames, like empty picture frames, that fit together, one on top the other, and are aligned by pins in one frame and holes in the other. The flask can be made of either wood or metal. Bronze flasks were in use by Biringuccio's time, and by the eighteenth century cast-iron ones were available, but undoubtedly wood was often employed. The inventory of the elder James Geddy's estate lists "founders flasks" but does not specify material.[24]

Patterns

In addition to the mold and the frame in which it will be constructed, one other element is essential. It is not practical, or even possible, to carve or sculpt directly in sand, so it is necessary to have a pattern, a model over which the sand can be packed to produce an imprint. In theory, a pattern can be made of any material that will withstand the packing process and that will not stick to the sand. In practice, patterns are almost always either wood or metal. Metal patterns in particular seem to have been standard in small works foundries, and a number were found on the Geddy site during archaeological work. In fact, in one rather remarkable instance archaeologists found a lead harness ring pattern on the Geddy property and a matching finished brass ring at another site in town, the James Wray property. Since the Wray example showed all the same minor idiosyncrasies of form, it must have been cast from the Geddy pattern.

Another pattern found on the Geddy site is of particular interest because it provides an example of pattern work in progress. This was a brass pattern for a shoe buckle, which was intended to have a small rosette design on each side, only one of which had been completely worked. Apparently the piece was lost before completion, or perhaps the customer canceled his order.

The archaeological survey also uncovered an example of the Geddys' use of an actual finished object as a pattern. A rough casting of a brass clock spandrel found on the site matched perfectly the spandrels used on a number of clocks produced in the Philadelphia area at a somewhat earlier period. It may be that someone in Williamsburg owned one of these Philadelphia clocks, that one of the spandrels was damaged or lost, and that the Geddys used one of the remaining spandrels to cast a replacement. This practice of casting from finished work naturally would be employed whenever it was

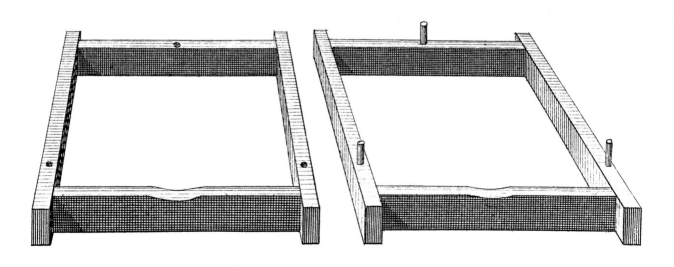

The two halves of a flask. Diderot,
Encyclopédie, *V, plate 3.*

feasible since it saved all the time and expense needed to produce a new pattern.

Patterns can often go through several generations before reaching finished form. The stem of a candlestick, for instance, might be first turned in wood on a lathe. The wood pattern would then be used to make a mold, which would be used to cast a permanent metal pattern. The making of patterns was an aspect of foundry work that later became a separate trade; by about 1820 references to pattern makers and pattern shops as separate entities started to appear. In the eighteenth century, however, pattern making was still an integral part of the founder's craft.

A pattern often must be more than just an exact model of the object to be made. Consideration must be given to the manner in which it will be used and the requirements of the molding and casting process. The exact form of a pattern will depend on a number of variables: whether the object to be made will be cast whole or in sections; whether the product will be solid or hollow; the vertical or horizontal disposition of the object in the mold; the size, shape, and direction of the channels by which the metal will be admitted into the mold cavity; the flow properties of the specific metal being cast; and several other factors. In addition, certain requirements or restraints are common to all patterns. Most metals will contract on cooling; therefore, where size of the finished article is critical, the pattern must be made oversize to compensate for the shrinkage of the metal. The pattern also should be of as uniform a thickness as possible in order to avoid localized shrinkage in heavier areas of the casting. Since the pattern must be removed from the mold before the metal is poured in, it must also have draft. If the sides of a pattern are undercut, or even if they are perfectly vertical, it cannot be removed, or drawn, without tearing the mold. Draft is simply a slight inclination away from the vertical that allows the pattern to be drawn without disturbing the sand that is packed against its sides.

At The Molding Bench

In actually forming a mold, the pattern initially is placed on a master mold, a bed of packed sand contained within a flask that positions the pattern so that the sand can be packed over it. A fine powder called parting, composed of talc, powdered charcoal, or similar materials, is dusted over the master mold and pattern to prevent sticking when the moist sand is packed over them. Half of an empty flask is fitted over the master mold. Sand is then

A pattern for a cooking pot and the mold made from it. A cooking pot is a type of larger object likely to have been cast in the Geddy foundry. Diderot, Encyclopédie, *IV, plate 7.*

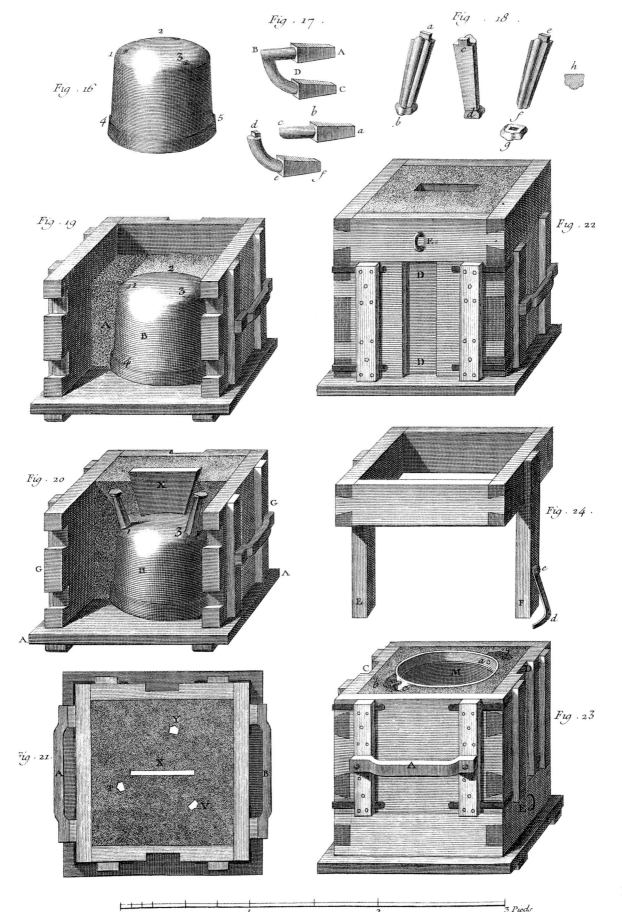

Fig. 16

Fig. 17

Fig. 18

Fig. 19

Fig. 20

Fig. 21

Fig. 22

Fig. 24

Fig. 23

1   2   3 Pieds

35

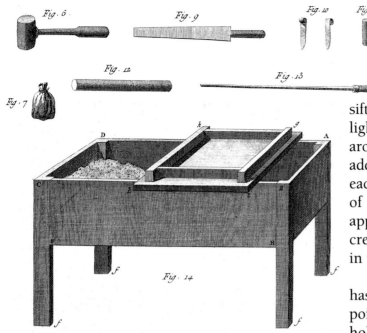

Fig. 6. Fig. 9. Fig. 10 Fig. 11

sifted into the flask and packed down lightly with the fingers over and around the pattern. More sand is added in stages, being packed down each time with a proper succession of tampers and mallets, the force applied in the packing gradually increasing with the depth of the sand in the flask.

When this first half of the mold has been packed, or rammed, to a point where it is just firm enough to hold its shape, it is lifted off the master and pattern and is sprayed lightly with water; a facing of ground soapstone, graphite, or other fine powder is dusted over it to provide a smooth, more finely detailed mold surface. The half-mold is then replaced on the master and is given a final packing to embed the facing particles in the sand and bring the entire half-mold to the desired firmness.

The whole thing—the master mold and the half-mold made on top of it—is then turned over and the master mold is carefully lifted off, revealing the pattern embedded in the half-mold that has just been formed. The other, empty half of the flask is placed over this half-mold and the ramming process is repeated to form the second half of the mold.

Having been made in halves, the mold now can be opened and the pattern removed. Channels to admit the metal, called sprues or gates, and vents to facilitate the escape of air and gases generated during the casting process are then carefully cut into the sand. The mold can now be closed up and baked dry to prepare for casting. Since each mold is good for only one pour, making molds is a never-ending process in the foundry.

For many shapes a two-part mold of the sort just described is all that is required. If a piece is to be cast hollow, however, the use of a core may be necessary. A core is simply a sand insert that is positioned inside the mold so that the metal will flow around it. When the casting is removed from the mold, the core is broken out leaving a hollow.

In British or British colonial foundries, though cores were common in many types of work, candlesticks and other thinner-walled objects tended to be cast in halves and brazed together to achieve a hollow. This practice continued until near the end of the eighteenth century, after which core

Molder's bench and tools: Figs. 6, 9, 12. Tools for packing sand. Fig. 7. Cloth bag for facing or parting powder. Figs. 10, 11. Tools for cutting sand to place patterns, form gates, etc. Fig. 13. Scraper to level back of mold. Fig. 14. Molding bench. Diderot, Encyclopédie, V, plate 2.

Counterclockwise from top—A wood pattern for a valve, the sand core, and the bronze casting of the valve. The metal flows around the sand core to make the casting hollow.

The molder has positioned the pattern for a candlestick base on the master mold; he now fits the top half of the flask over the master mold.

After dusting the master mold with parting powder, sand is sifted into the flask.

The sand is packed down in stages—first with the fingers, then tampers, and finally a mallet.

Once the sand in the first half of the mold has been packed tightly enough to hold together, it is lifted away from the master mold. The half-mold will then be dusted with a facing powder and replaced on the master mold for a final packing.

After making both halves of the mold, the molder removes the pattern; he carefully cuts sprues, or gates, into the sand to admit the molten metal into the mold.

The left half of the completed mold shows the sprues and also the vents that will facilitate the escape of air and gases. The vents are the grooves radiating to the edge of the mold.

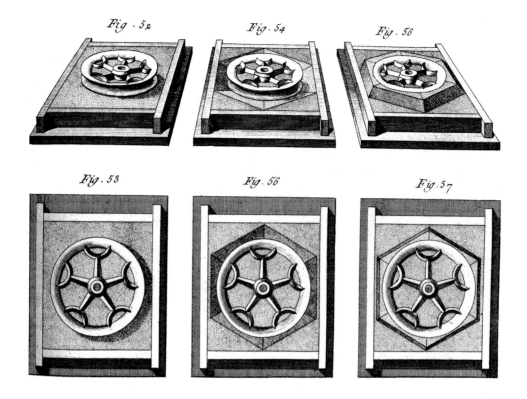

casting became standard in English shops as well. Consequently, candlesticks and similar forms produced in England or British America before 1800 can often be identified by the solder seam that is almost always visible on close examination.

Another class of objects that cannot be produced in a two-part mold includes those that have undercuts as part of their design. The pulley wheel shown in the illustration is a good example of this. The problem—the undercut—in this case is the groove running along the rim of the wheel. In attempting to make this with a two-part mold, the sand that is packed over the wheel would also be packed into the groove, and there would be no way to remove the pattern from the sand without tearing the mold. The solution can be seen in the last part of the illustration. Independent bodies of sand called false cores are formed along the undercut area. These may be removed to allow the pattern to be drawn and then replaced when the mold is assembled for pouring.

This false core work is perhaps the most highly skilled aspect of molding, as molding is perhaps the most technically demanding aspect of the founder's trade. It takes years of training and practice to become an expert molder, one with the "molder's touch," a sensitivity to subtle gradations in the condition of the sand and the increasing firmness of the

Mold with false cores. Diderot,
Encyclopédie, V, plate 6.

mold under the packing process; one with the ability to effect quickly and with a minimum of forethought the constant necessary manipulations and adjustments of pattern and mold surface; and one with the capacity to conceptualize quickly and accurately any three-dimensional form in terms of volumes and surfaces that must be reproduced in reverse in the molding medium in a manner that will allow the removal of the pattern and the subsequent introduction of the metal in the most effective fashion.

Speed counts. As soon as the sand begins to be worked, it begins to dry. Frequent and precisely regulated applications of water to the mold surface can only retard the loss of temper, and the sand gradually loses its plasticity as the moisture level drops. The molding process must be careful and deliberate, but it cannot be leisured.

In William Geddy's shop another consideration also demanded quickness in execution. Quicker work meant greater profits. Whether John Dennis could make three molds in an hour or six was not a matter of indifference. Production capacity was wealth. This imperative for efficiency applied not just to molding, of course, but to all aspects of the work and to all members of the staff. Young Thomas Dugar's labor was cheap compared to that of a highly paid journeyman like Dennis, but it was not free. He received no wages at the end of the week, but he ate food and wore clothes every day. Many of the jobs he was assigned were semiskilled or unskilled, but none of them was unnecessary to the operation of the shop, and inefficiency, even in an apprentice, could be a serious drag on production. The craft education of an apprentice like Dugar was not only a matter of "how to" but also "how many."

Perhaps our mental picture of the traditional craftsman is too often colored by our image of an older way of life and work that was somehow more leisured and stable, less frenzied and uncertain than our own. We look at the antique work and say: "Imagine how much time it must have taken to make that. But, then, they had plenty of time in those days." The image becomes vaguely idyllic; we see the cloistered craftsman, tradesman as artist, sequestered in the tranquillity of his quiet studio, slowly and painstakingly applying the finishing touches to his latest masterpiece. Imagine, instead, John Dennis at the molding bench in William Geddy's shop on a hot, sticky August afternoon. To be sure, the work he is doing requires great skill, even artistry if you will. But look at him again as he stands at the bench pounding sand, working in a cloud of dust, skin covered with a layer of sand, talc, charcoal, and graphite, bonded with sweat; a 2000° fire in the forge eight feet

from his back; seven hours into the workday and only halfway through; a mold and a half behind the pace he needs to maintain to get the day's work out, and the sand somehow is not working right: it feels gritty—wasn't tewed properly—Dugar can't even do that right. Seeing Dennis at this moment, we remember that the craftsman is an artist occasionally, a worker always.

It has been said that a molder probably reaches peak production capacity in his forties, when he has had time to acquire all the hundreds of tricks of the trade but still retains most of his physical strength and endurance. This is certainly debatable on many counts, but it does suggest the combination of technical and physical demands that the process involves. It also suggests the consideration that when a shop master such as William Geddy hired an experienced craftsman like John Dennis, he acquired not only an able pair of hands but also many of the trade secrets of any shop in which Dennis had worked previously. The movement of skilled journeymen from shop to shop undoubtedly played a key role in the dissemination of technology in a time when technical publications were almost nonexistent and the mysteries of the craft were jealously guarded.

At The Forge

When the molds have been made and baked dry, they are stacked in a press, clamped together, and turned on end for pouring in front of the forge. Since this forge is used for melting a substantial amount of metal (ten to twenty pounds on average), it differs somewhat from the typical blacksmith's forge, which is used to heat a relatively small section of iron without actually reducing it to a liquid. Since a crucible containing the metal must be set all the way down in the coals, with a sufficient bed of fuel remaining beneath, the foundry forge tends to be considerably deeper than it is broad. The forge found on the Geddy site

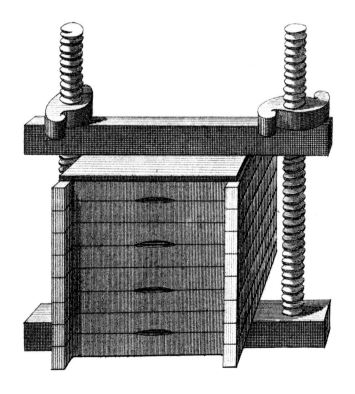

Left: The molten metal is poured into several molds at a time; the molds have been baked dry, stacked in a press, clamped together, and turned on end in front of the forge. Right: Sand molds in press ready for pouring. Diderot, Encyclopédie, V, plate 6.

appears to have been a brick pit about one foot square and two and one-half to three feet deep with an iron grate about a foot from the bottom to allow ashes to fall through. A tuyere, or pipe, from a hand-operated bellows entered from the side of the forge toward the bottom, and a constant draft would have been maintained during the entire course of the melt, which might have lasted an hour or more. Needless to say, the operation of the bellows would have been delegated to one of the apprentices in the shop.

The conditions under which the metal is melted and the temperature at which it is poured are critical to the success of the casting. At high temperatures most metals will oxidize rapidly if exposed to air, so the crucible is covered with a lid and often a piece of charcoal is placed on the surface of the molten metal to consume oxygen before it can enter the melt. Exposure to air can cause excessive loss of metal in the form of slag, oxidized metal that floats up and forms a layer of scum on the surface of the melt. The presence of oxides in the melt also greatly reduces the fluidity of the metal when it is finally poured.

Gases other than oxygen can also create difficulties. Brass, bronze, and silver, which were the staple metals of the Geddy foundry, have a tendency to absorb hydrogen from the combustion of the fuel and from water vapor in the atmosphere. This hydrogen is dissolved in the metal as long as it is molten, but it is driven off as vapor once again when the metal starts to solidify after casting. Some of this gas can be absorbed by the permeable wall of the mold and some by the metal in the gates, which is the last to solidify. But, if an excessive amount is present in the melt, it will cause porosity in the casting. Ironically, the most effective way available in the eighteenth century to de-gas a melt was to let the metal oxidize to some extent, which would draw off hydrogen as the oxides floated to the surface of the metal. Thus, the craftsman had to develop a finely tuned sense of the state of the melt and the atmosphere in his forge, balancing between oxidizing and reducing environments to achieve optimum conditions for the pour. This rather complex technical expertise is particularly striking when we consider that the eighteenth-century founder had no theoretical knowledge of the actual chemical mechanism just described. Through his apprenticeship he had to learn to observe subtle changes in the appearance of the metal, recognizing which boded ill and which good, and to adjust the conditions in forge and crucible accordingly.

In the same manner the temperature of the metal had to be determined by eye, the craftsman observing closely the color of the crucible and the fire,

Opening sand molds after pouring. Sand from the mold can be seen clinging to the casting. The sand is easily brushed from the casting, but the mold is ruined.

*The rough castings for a candlestick
with the sprues still attached.*

the movement of convection currents in the metal, and other indicators of relative heat. Brass, for instance, an alloy of copper and zinc, will start to show an actual boiling of the metal and a bright white flame and whitish smoke coming off the surface as it approaches pouring temperature and the zinc starts to vaporize. Adding to the difficulty of accurate judgment is the fact that the desirable pouring temperature for brass or bronze is not a constant. Fairly slight differences in the composition of the alloy can significantly affect the temperature requirements. The size and shape of the article being cast and the design of the mold's feeding system also have to be considered in determining the proper time to pour.

When the metal is finally judged to be ready, the craftsman lifts the crucible from the forge with iron tongs and pours into the molds. The castings are allowed to cool in the molds for perhaps five minutes to one-half hour, depending on the size of the piece. The molds are then opened and the castings removed. At this point the craftsman will have some idea whether this metamorphosis of matter from solid to liquid and back to solid that he has wrought has produced a collection of objects of use, beauty, and value or whether he has simply transformed twenty pounds of scrap metal into twenty pounds of scrap metal.

Biringuccio's observations on the uncertainty and anxiety inherent in the founder's trade are all too true, and a score of different types of casting flaws can occur for which there is no remedy but remelting. Of course, all these defects must be caused by something that was poorly conceived or improperly executed in the molding or casting process, and it is the business of the founder to foresee these problems and allow no chance for them to occur. Perhaps Biringuccio said it best: "To conclude: The outcome of this art is dependent upon and subject to many operations which, if they are not all carried out with great care and diligence and well observed throughout, convert the whole into nothing, and the result becomes like its name [cast away]."[25]

Finishing

Though a great deal of technical skill and not inconsiderable labor have been expended up to this point in order to construct the mold and pour the metal, the subsequent finishing of the rough casting consumes the bulk of the time required to produce a given article. First, any sand clinging to the casting is brushed off and the actual piece is cut away from the gates. Then

the surface is tooled in some fashion—edges and larger surfaces are filed or scraped; detailed work is smoothed and defined with chasing tools, small blunt-ended steel punches worked with a hammer; and round or cylindrical forms are skimmed with cutting tools on a hand- or foot-powered lathe. The actual polishing is done by hand rubbing the tooled surface with ground stone abrasives applied with a rag or piece of leather. A series of such abrasives are used in sequence, from coarser to finer, and might include emery, pumice, tripoli, and rouge, all of which are naturally occurring rock materials and were sold in Williamsburg stores in the eighteenth century. Some objects, of course, are cast in pieces rather than entire, and in that case the sections have to be fitted and assembled.

Almost certainly apprentices did the greatest part of the finishing work. They were the cheapest labor in the shop, receiving no cash wages at all, and it would not make economic sense for the highly paid journeyman craftsman to perform work that an apprentice could do just as well. Young Thomas Dugar undoubtedly spent most of his first year in William Geddy's shop polishing castings, pumping the bellows, bringing in coal for the forge, cutting up scrap metal, breaking up used sand molds, tewing the sand, sweeping the floor, and performing innumerable other less skilled, dirty, and often boring tasks.

His position in the shop was not static, however. Just as he was under

Above: The rough-cast candlestick base is filed, preliminary to polishing. Right: The high luster of the finished piece is obtained by polishing with a series of fine abrasives.

a legal obligation to perform faithfully the work that William assigned him, the same contract of apprenticeship bound William to teach Dugar all the mysteries of the trade. As the months and years went by, as Dugar grew from a boy to a young man, he would gradually come to participate in all aspects of the craft work. Although in a legal sense his status remained the same for the whole period of his service, his importance to the shop would grow with his craft skills. By the last year of his apprenticeship he would probably feel with some justification that he was no longer just an apprentice boy but was truly a colleague—although a junior one—of the old hands like John Dennis.

That first year in service, though, must have been a memorable one for Dugar or any other apprentice. Consider for a moment a young boy, maybe only ten or eleven years old, leaving the home and people he has known; going to live with a new family; finding himself thrust into the dirt and din of a shop where most of the activities, indeed the whole rhythm of the workday, are largely mysterious; spending most of his waking moments, twelve or fourteen hours a day, six days a week, in these strange surroundings performing many dull, laborious, and—to his untrained hands—difficult duties under the watchful eyes of a journeyman or the master himself, who may or may not sympathize with his inexperience, but who most certainly knows that time is money—even an apprentice's time.

At least Dugar did not find himself alone in his plight. There was, as noted earlier, at least one other apprentice in William Geddy's shop, and he was perhaps a particularly useful ally since he was Geddy's own son, William, Jr. William, Jr., would undoubtedly have had a wealth of valuable information to pass on concerning the workings of the foundry and, perhaps even more valuable, the idiosyncrasies of the old man. It should be remembered, in any case, that the apprentice normally lived with the craftsman he served and was, in effect, a member of the family. If the master was a boss, he was also a mentor. If the workshop was a place of drudgery, it was also a place of learning; the long hours at the forge and bench were repaid with a growing sense of competence, self-reliance, and gradual unfolding of creative potential. The bonds of apprenticeship were not necessarily only contractual obligations but could also be bonds of affection and mutual respect.

The finished candlestick with its component parts.

The Metals

Most of the articles produced in the Geddy foundry were made of brass or bronze. Both of these are copper-based alloys, brass being a mixture of copper and zinc and bronze of copper and tin. Actually, through most of the eighteenth century, the term *brass* was used indiscriminately for any copper alloy; the word *bronze* was just beginning at that time to come into  use to distinguish the copper–tin from the copper–zinc combination. Eighteenth-century documents often refer to brass cannons, bells, armor, or statuary, all of which would be made normally of what we today call bronze.

Bronze is much the older of the two metals, being the first alloy developed by man. Brass, on the other hand, seems not to have come into really wide use until the Middle Ages. This was undoubtedly due to the difficulty of making the alloy. The problem was the vaporization point of zinc, which is unusually low, lower, in fact, than the temperature required to extract zinc from its ore. This meant that attempts to remove the metallic zinc from calamine, or zinc ore, simply resulted in boiling off all the metal into the atmosphere. Consequently, brass had to be made by a laborious and difficult to regulate process of layering charcoal, zinc ore, and copper fragments in a closed container, heating this for a long period, and allowing the copper to absorb the zinc vapor. Often this was thought of as a process of coloring copper rather than actually mixing metals.

This remained the state of things well into the eighteenth century. A method of refining zinc by a sort of distillation process was finally developed and patented in the 1730s, but this method did not come into wide use until the very end of the century. Oddly enough, actual refined metallic zinc had been available for some time as an import from the Orient, but the relationship of this spelter, as it was called, and the calamine ore was not understood in the West, and consequently it was not used to produce brass. Brass, nonetheless, had gained steadily in popularity and by the eighteenth

Above: Casting pewter. Smiths and founders often worked in pewter as a sideline. "L'Art du Potier D'étain," Descriptions des Arts et des Métiers, Vol. XXII (1774; Slatkine Reprints, Geneva, 1984), plate 28. Right: Pouring a pewter spoon.

century was definitely more common than bronze.

Bronze, however, was used more often for domestic articles than it is today, and it was not unusual to see a bronze candlestick or andiron. In some cases, such as making cannons or bells, bronze might be used rather than brass because of its greater strength or hardness.

As was mentioned earlier, the foundry on the Geddy property was also casting precious metals. Although silver was generally hammered to shape, a fairly large class of objects—usually smaller, very ornate pieces—was commonly cast. A good bit of silver casting on the site was likely for the production of trim pieces for larger hammered vessels. A coffeepot, for instance, though essentially a hammered form, has a spout, finial, and sockets for the handle that are cast. The body of a tray is hammered; its borders and feet are cast. Although relatively little gold was worked in British America, apparently it too was cast on the Geddy property, since fragments of two crucibles were found with traces of gold stuck to the inside.

One other metal worked by the Geddys, at least in the earlier years, was pewter. Pewter was the most widely used of all nonferrous metals. It is a variable tin-based alloy—in the case of fine pewter usually around 90 percent tin—with copper and antimony added in small quantities as hardening agents. Cheaper grades would have somewhat less tin and sometimes lead added as an adulterant or to enhance the casting qualities of the alloy. The process for casting pewter is radically different from that for the other metals. Since it has a much lower melting temperature, it can be poured into permanent molds made of bronze rather than sand. The casting of pewter was, in fact, a separate trade, that of the pewterer, but smiths and founders often worked in pewter as a sideline. There seems to have been a particularly good market for pewter spoons. The inventory of Adam Pavey's estate lists nine hundred pewter spoons along with his blacksmith's wares, and James Geddy, Sr.'s inventory lists four spoon molds and one hundred pounds of scrap pewter.[26]

Geddys After 1779

The long history of metalworking on the Geddy site finally ended in 1779 when James Geddy, Jr., sold the property to Robert Jackson, a local merchant. James moved to a farm in Dinwiddie County for a few years, though he seems to have continued doing some silversmithing during this period. By 1783 he had once again opened a shop in the nearby town of

Petersburg, and shortly thereafter he went into partnership with his sons.[27]

James's sale of the Williamsburg property had, of course, also displaced his brother William's foundry, but William remained in the area until his death in 1783. His son, William, Jr., moved to Richmond and went into business there, as did William Waddill, the Geddys' brother-in-law and former business associate.[28] During this period many other craftsmen were also leaving Williamsburg, reflecting a general westward shift of population and the move of the capital of Virginia to Richmond. Williamsburg was becoming a much less attractive place for tradesmen.

The last of the Geddy craftsmen to work in Williamsburg seems to have been the slave Charles who was mentioned earlier and who appears in the following statement of a Richmond justice of the peace:

> City of Richmond to Wit
>
> This day George P. Crump of the city aforesaid personally appeared before me a justice of the peace . . . and made and subscribed the following statement to Wit: The said George P. Crump says that he has been summoned to appear before the Hustings Court of the City of Williamsburg to answer a presentment in said court against him for permitting an old negro man (a blacksmith) named Charles to go at large and hire himself out in the City of Williamsburg.
>
> The said negro Charles was the property of old Mr. William Geddy who died about the year 1817: By the will of Mr. Geddy Charles was loaned to his widow during her life and at her death was to be free. . . . The widow of Wm. Geddy was the sister of your affiant, and has been dead about five years.
>
> Geo. P. Crump
>
> Sworn and subscribed before . . . this 26th day of May 1843
> Walter D. Blair[29]

Notes

[1] Vannoccio Biringuccio, *Pirotechnia,* trans. Cyril Stanley Smith and Martha Teach Gnudi (Cambridge, Mass., 1942), pp. 213–214.

[2] *Ibid.,* p. 72.

[3] Hunter's *Virginia Gazette* (Williamsburg), Aug. 8, 1751.

[4] Henry Hamilton, *The English Brass and Copper Industries to 1800,* 2nd ed. (New York, 1967), p. 264.

[5] For 1733 court records, see York County, Virginia, Records, Wills and Inventories, No. XVIII, 1732–1740, p. 68, microfilm, Colonial Williamsburg Foundation Library; for 1738 residence in Williamsburg, see York Co. Recs., Deeds, No. IV, 1729–1740, pp. 535–536, microfilm, Foundation Lib.; for advertisements, see Parks's *Va. Gaz.* (Williamsburg), Oct. 6, 1738, and Oct. 5, 1739.

[6] York Co. Recs., Wills and Inventories, No. XIX, 1740–1746, pp. 321–323, microfilm, Foundation Lib.

[7] *Ibid.,* pp. 21, 42.

[8] *Ibid.,* pp. 293, 300.

[9] For James Geddy's will, see *ibid.,* pp. 306–307; for Anne Geddy's petition, see H. R. McIlwaine, ed., *Journals of the House of Burgesses, 1742–1749* (Richmond, Va., 1909), p. 118; for award of payment, see Wilmer L. Hall, ed., *Executive Journals of the Council of Colonial Virginia, November 1, 1739–May 7, 1754* (Richmond, Va., 1945), p. 174; for William Beadle, see York Co. Recs., Wills and Inventories, No. XIX, pp. 321–323.

[10] Spotsylvania County, Virginia, Records, Minute Book, 1755–1765, p. 41.

[11] For Galt, see *Va. Gaz.* (Hunter), Sept. 2, 1757; for James, see York Co. Recs., Deeds, No. VI, 1755–1763, pp. 276–278, microfilm, Foundation Lib.; for John, see Purdie and Dixon's *Va. Gaz.* (Williamsburg), July 29, 1773.

[12] Spotsylvania Co. Recs., Orders, 1738–1739, pp. 464, 469, 497.

[13] For purchase, see York Co. Recs., Deeds, No. VI, pp. 276–278; for rebuilding, see Herman J. Heikkenen and Peter J. J. Egan, *The Years of Construction for Eight Historical Structures in Colonial Williamsburg, Virginia, as Derived by the Key-Year Dendrochronology Technique* (Blacksburg, Va., 1984).

[14] For Goosley's ad, see *Va. Gaz.* (Hunter), June 6, 13, 20, 1751; for Haldane's ad, see Dixon and Hunter's *Va. Gaz.* (Williamsburg), July 13, 1776.

[15] *Ibid.* (Purdie and Dixon), Aug. 19, 1773.

[16] *Ibid.* (Hunter), Oct. 17, 1751.

[17] *Ibid.* (Dixon and Hunter), July 13, 1776.

[18] Records of the Public Store in Williamsburg, 1775–1780, Journal, Sept. 14, 1778–Nov. 30, 1779, p. 269, Virginia State Library, Richmond, microfilm, Foundation Lib.

[19] For William Geddy's town residence, see "A List of Taxable Articles in the City of Williamsburg Taken by Robert Nicolson for the Year 1783 under the Revenue Act," *William and Mary Quarterly,* 1st Ser., XXIII (1914), p. 136; for both William and Anne Geddy's rural properties, see Williamsburg–James City County, Virginia, Tax Book, 1768–1769.

[20] C. G. Chamberlayne, ed., *The Vestry Book of Blisland (Blissland) Parish, New Kent and James City Counties, Virginia, 1721–1786* (Richmond, Va., 1935; reprinted, 1979), p. 161.

[21] For Dennis at Tayloe Ironworks, see Tayloe Account Book, 1740–1741, Tayloe Papers, Virginia Historical Society, Richmond; for Dennis's inventory, see Halifax County, Virginia, Will Book I, 1773–1782, pp. 279–280.

[22] Biringuccio, *Pirotechnia,* p. 214.

[23] *Encyclopaedia Britannica,* 1st ed., s.v. "foundery."

[24] York Co. Recs., Wills and Inventories, No. XIX, pp. 321–323.

[25] Biringuccio, *Pirotechnia,* p. 213.

[26] For Pavey's estate, see Spotsylvania Co. Recs., Will Book B, 1749–1759, pp. 293–294; for Geddy's inventory, see York Co. Recs., Wills and Inventories, No. XIX, pp. 321–323.

[27] For sale of property, see Purdie's *Va. Gaz.* (Williamsburg), May 2, 1777; for move to farm, see York Co. Recs., Deeds, No. VI, 1777–1791, pp. 48–49, microfilm, Foundation Lib.; for shop in Petersburg, see *Virginia Gazette, or, the American Advertiser* (Richmond), Aug. 16, 1783.

[28] For William's death, see James City County, Virginia, Land Tax Lists, 1782–1805; for William, Jr.'s move to Richmond, see Wilmer L. Hall, ed., *Journals of the Council of the State of Virginia, (December 1, 1781–November 29, 1786),* III (Richmond, Va., 1952), pp. 387ff.; for Waddill's move to Richmond, see *Va. Gaz., or, Amer. Ad.,* Apr. 23, 1785.

[29] Southall Family Papers, 1807–1904, Folder 147, Manuscripts and Rare Books Department, Swem Library, College of William and Mary, Williamsburg, Va.

The authors would like to thank Harold B. Gill, Jr., for researching and collecting the primary documents relating to the Geddy family.

Further Reading

Biringuccio, Vannoccio. *Pirotechnia*. Translated by Cyril Stanley Smith and Martha Teach Gnudi. Cambridge, Mass.: M. I. T. Press, 1942.

Gentle, Rupert, and Rachael Feild. *English Domestic Brass, 1680–1810, and the History of Its Origins*. New York: E. P. Dutton and Co., 1975.

Hamilton, Henry. *The English Brass and Copper Industries to 1800*. 2nd ed. London: Frank Cass and Company, 1967.

Montgomery, Charles F. *A History of American Pewter*. New York: Praeger Publishers, 1973.

Noël Hume, Ivor. *James Geddy and Sons: Colonial Craftsmen*. Williamsburg, Va.: The Colonial Williamsburg Foundation, 1970.

Raymond, Robert. *Out of the Fiery Furnace: The Impact of Metals on the History of Mankind*. University Park, Pa.: Pennsylvania State University Press, 1986.

Tylecote, R. F. *Metallurgy in Archaeology: A Prehistory of Metallurgy in the British Isles*. London: Edward Arnold, 1962.